COVID-19

A PHYSICIAN'S TAKE ON THE EXAGGERATED FEAR OF THE CORONAVIRUS

5TH EDITION

$12.95

COVID-19

A PHYSICIAN'S TAKE ON THE EXAGGERATED FEAR OF THE CORONAVIRUS

5TH EDITION

Jeffrey I. Barke, M.D.

THE AMERICAS GROUP

FIRST EDITION
1st Printing — August 2020
2nd Printing — October 2020
SECOND EDITION
1st Printing — November 2020
THIRD EDITION
1st Printing — January 2021
FOURTH EDITION
1st Printing — August 2021
2nd Printing — September 2021
3rd Printing — October 2021
4th Printing — November 2021
FIFTH EDITION
1st Printing — January 2022

The Americas Group
520 S. Sepulveda Blvd., Suite 204
Los Angeles, California 90049-3534 U.S.A.

ISBN

978-0-935047-97-4

Description: Second edition. | Los Angeles, CA : The Americas Group, 2020. | Includes index. | Summary: "This book of brief essays methodically examines the major Covid-19 issues to determine what is fact and what is myth. It is written by a family physician who pointedly differs on the threat posed by Covid-19 to the American people, then challenges current conventional wisdom on what should be done about the schools, the economy and public activity with reason, honesty and scientific findings."— Provided by publisher.

Printed in India by
AEGEAN OFFSET PRINTING, LTD.

CONTENTS

Foreword 7

Introduction 11

First, Do No Harm —
  *Premium non
  nocere* 13

Immune Systems
  Matter 19

Size Matters in
  Science and
  Common Sense 25

Better To Be Good
  or Lucky? 33

Not Dying Is Bad for
  Headlines — Covid-19
  Cases v. Death 41

She Died Alone 46

Notes/Questions 52

Boosters   53

Put Your
   Damn Mask On   61

Here We Go Again —
   Covid 19/3.0   67

It's Just a Test   75

Fear Fatique   83

The Covid-19
   Vaccines: Risks
   and Benefits   89

Concluding Thoughts
   on Covid-19   105

About the Author   107

Acknowledgments   109

Index   113

FOREWORD

THE TALMUD, JU-DAISM'S second holiest work after the Bible, a repository of law, folklore and wisdom the size of an encyclopedia, contains an amazing statement: "The best physicians go to hell."

Now, why would a holy work say such a thing about physicians, those who are arguably engaged in the holiest work of all—saving human lives? Moreover, why would it say such a thing about "the best" of them?

The answer is the ancient rabbis understood how tempting it can be for a physician, more than for the members of other professions, to think of himself — and for others to think of him — as a god.

I can say with certainty that the Talmudic principle does not apply to Dr. Jeffrey Barke. He is certainly one of the best physicians, but he knows that no doctor, not even those who run the CDC or the FDA or who edit the *New England Journal of Medicine* or *The Lancet*, is a god. That is why he is prepared to challenge conventional medical wisdom when doing so is warranted by the facts.

He has taken a hard look at the lockdowns implemented in America and in almost every other country at the behest of doctors in response to the novel coronavirus known as Covid-19, and he sees a mistake the likes of which no one alive at this time has seen.

Unlike most of his colleagues, he not only sees the price people pay because of the virus, he sees the price paid because of the lockdown: the impoverishment of hundreds of millions of people around the world, the impoverishment and near-impoverishment of countless Americans, the increase in depression, suicide, children's loss of education, family tension, increased drug use, recovering addicts returning to their addiction, people delaying or forgoing necessary medical treatments and much more. And he sees the political turmoil that inactivity, economic depression and loss of income inevitably lead to.

Also, unlike many of his colleagues, he has the courage to advocate medical treatments that work, irrespective of what the medical establishment has pronounced, not because of science, but because of politics.

The combination of medical expertise, courage, and wisdom is very rare. We have all three traits in Jeffrey Barke. That's why he needs to be heard.

And that's why I think the ancient rabbis would have said, "Here's one physician who will go to Heaven."

I certainly think so.

Dennis Prager
Radio talk show host,
lecturer and author
whose intellect and integrity
have influenced millions of
lives

INTRODUCTION

THERE ARE MANY great traditions in American political life. Freedom of speech is perhaps the most fundamental. That freedom not only permits the widest possible expression of views but encourages dissent no matter how broadly or how firmly the majority view is held.

Dealing with Covid-19 is one of those subjects that is stimulating continuing controversy. This fourth printing of the fourth edition of *A Physician's Take on the Exaggerated Fear of the Coronavirus* allows Dr. Barke to refine his thinking and add new thoughts since publication of the first edition.

Just as the four physicians who signed the American Declaration of Independence pledged their lives, their fortunes and their sacred honor in the renunciation of the rule of King George III in the American colonies, so Dr. Barke, a family physician in Corona del

Mar, California, differs with the conventional wisdom concerning the threat of Covid-19 to the American population.

In the essays on the following pages, Dr. Barke examines the major Covid-19 issues to determine what is fact and what is myth. He buttresses his analyses with reason, honesty and scientific findings that go against much of what the public has been led to believe from government officials and the media.

Readers are urged to examine for themselves whether his ideas make sense. Keep an open mind. Be part of that other great American tradition that believes in unbridled and fair debate to arrive at fresh conclusions. Remember the words of George Bernard Shaw: "Those who cannot change their minds, cannot change anything."

Godfrey Harris
Public policy consultant
and managing editor of
The Americas Group

FIRST, DO NO HARM —
PRIMUM NON NOCERE

HIPPOCRATES WAS A Greek physician who lived some 400 years before the common era. He is often referred to as the father of modern medicine because he was one of the first to describe his medical observations in a scientific manner. He wrote more than 70 books.

He is perhaps best known for his dictum expessed in Latin: *Primum non nocere* — or, "First, do no harm." It is part of the Hippocratic oath that is still attested to by medical school graduates.

Politicians should be

administered a similar oath when they take office. If they were, they might not have done all the questionable things we are dealing with in the current Covid-19 pandemic.

There is no doubt that Covid-19 is a dangerous virus to the elderly and the frail. But the fact is that our *reaction* to the virus has caused more harm than the virus itself.

As more and more data become available, it is clear that the resulting fatality rate from this virus will be around 0.2%. That is in the ballpark of a bad influenza season. It is also twenty times lower than originally assumed by the World Health Organization (WHO).

The fatality rate for Covid-19 residents of nursing homes and assisted living facilities account for close to 50% of all deaths in the United States; younger Americans have a much lower fatality rate from this disease.

In fact, the average age of Covid's fatal victims in most

countries is more than 80. For Americans under the age of 25 there is a greater risk of being killed in an automobile accident than dying from Covid-19. Despite this, we have closed our schools and continue to cogitate about how and when to reopen — and in some cases — keep them open.

Moreover, the Covid-19 death toll itself is also now coming into question as the official figures do not differenti-ate death *from* Covid-19 versus death *with* Covid-19.

In addition, we are learn-ing that many presumed Covid-19 deaths had no laboratory confirmation and may have been coded as such to capture increased reimbursement from various government agencies.

I believe that when the history books are written about this pandemic, they will show that our reaction to this virus was a great mistake. Worse, the continuation of our initial response to the virus is no longer just a mistake, it is bordering on malice.

With more than 30 million Americans unemployed at the height of the initial reaction to the spread of the disease and with estimates that 40% of those will not have jobs to return to, the economic devastation initially crippled America and the world. This self-inflicted economic wound has also caused an almost complete shutdown of the U.S. outpatient healthcare system.

Doctors from all over the country report devastating consequences to their patients. Recently I co-authored with Simone Gold, M.D., J.D., a letter to President Trump highlighting this aspect of the healthcare crisis. The letter can be viewed at *www.adoctoraday.com*.

Two days after the White House received this letter, Dr. Anthony Fauci of the President's Coronavirus Task Force changed his message to the public:

> *Stay-at-home orders intended to curb the spread of the coronavirus could cause "irreparable damage."*

A psychiatrist reported that his office volume is down by 80% and yet his prescriptions for benzodiazepines (Valium, Xanax, Ativan) are the highest in his career. Patients who require frequent visits to stay functional are not coming into his office for fear of contracting the virus.

A cardiologist reports that routine echocardiograms and other non-invasive cardiac testing protocols are not being administered because these procedures have been declared non-essential. Patients are afraid to visit a doctor's office.

A gastroenterologist reported that routine cancer screening colonoscopies have stopped altogether, and a gynecologist reported that cancer-screening pap smears and mammograms have ceased.

I am personally aware of a

61-year-old patient who died of an intestinal obstruction because he feared going into his doctor's office. Instead, he suffered at home, ultimately succumbing to sepsis.

These examples are being repeated across the country with devastating consequences. False information about waiting for a vaccination, the need for widespread testing, mask-wearing and potential asymptomatic spread of Covid-19 fill the airways and stoke unnecessary fear among the American citizenry.

Winston Churchill once noted that "Fear is a reaction, courage a decision." Maybe it is time for us to remember Dr. Hippocrates' statement from long ago: "First, do no harm."

It would go a long way to putting the current situation on the right track, not just by our doctors but by our elected officials as well.

IMMUNE SYSTEMS MATTER

PLANS AND DEMANDS for school reopenings are coming in fast and furious from government and nongovernment organizations alike. Some are hundreds of pages long, requiring a phalanx of Ph.D.s to sort through the details before implementation.

What seems universally clear is that no one is taking into account that the vast majority of us have immensely powerful immune systems that play a critical role in keeping us healthy and alive.

One plan calls for using sanitizing spray on all classroom surfaces multiple times per day; that children use hand sanitizer upon entering and exiting the classroom; and that all children and staff wear masks for the entire school day and, of course, practice social distancing. The bureaucrats at the CDC would be proud of the results.

But all of these "specialists" have ignored the fact that from the day we are born we are assaulted by germs — by the millions, if not billions. Our very existence is dependent on a robust immune system — that is, the ability of our bodies to fight off any invasion of bacteria, viruses, funguses, molds and other pathogens. Fortunately, we were created with a powerful internal standing army of cells ready to protect us in each battle and capable of winning most wars.

In order for our immune system to be prepared for those battles, it has to train regularly and bring new recruits to the effort. Even before we are born, our immune system is exposed to germs and is working to protect us.

The result is that our bodies create germ-specific special ops fighters to defeat a variety of enemies that life throws at us. Each time we are exposed to new or old germs, our immune systems grow smarter and stronger.

It is healthy and necessary for our very survival to be exposed to different germs and to recover to fight another day. If we purposely prevent such exposure, we may gain in the short term, but we may also lose in devastating ways in the long term.

You may remember seeing on TV an episode of *Seinfeld* titled "The Bubble Boy." In this 30-minute story a boy needed to live inside a plastic bubble because he did not have a functioning immune system. It didn't end well for the "Bubble Boy" when his germ-free bubble was violated and he was exposed to germs.

Modern society has gone overboard with deploying antibacterial soaps, lotions and cleaning products. They indiscriminately kill germs, yes, but they also wipe out the good bacteria that help maintain a strong and diverse microbiome. "Kills up to 99.9% of common germs," promises the label of one brand of hand sanitizer.

Everyone has a microbiome, a collection of more than 100 trillion (!) microbes that live on and in our bodies, the majority in our large (and clearly crowded) intestine. The more diverse your individual microbiome, the healthier you are.

Research indicates that early exposure to a variety of microbes may help lower the risk of developing conditions like asthma, allergies and even infectious diseases. Think of it this way: If you exercise regularly and your body is fit, you are less likely to be injured, be overweight, have cardiovascular disease or suffer from diabetes. When you stop exercising, your level of fitness declines, along with all the benefits.

Your immune system works in the same way: Stimulate it regularly, such as when a child plays in the dirt, and you are more likely to win the battles against dangerous germs and viruses, including Covid-19.

With Covid-19, we have gone "Bubble Boy" sterilization crazy, and it is not helping us. We now sanitize everything: bus seats, door handles, gas pumps, purchased products, our bodies.

We are cleaning our homes (and some offices) as if they were an extension of a hospital's ICU. I have one patient who told me he comes home from the market, takes off all of his clothes outside, hoses himself off, puts the clothes in a bag, and then takes a hot shower.

The fear of Covid-19 has driven us to sometimes ridiculous and unhealthful behavior.

SIZE MATTERS IN SCIENCE AND COMMON SENSE

THIS ESSAY HAS BEEN UPDATED AND EXPANDED.
IN ORDER TO ACCOMMODATE IT TO THE SPACE
AVAILABLE, THE TYPE SIZE HAS BEEN REDUCED.

GOVERNOR GAVIN NEWSOM ISSUED AN EDICT ON June 18, 2020 that all citizens of California must wear masks in public to protect against the spread of Covid-19. A lively debate immediately ensued whether he had the authority to issue such an order. And more importantly, even if he had the authority, does mandating the wearing of masks make any common or scientific sense?

Early in the pandemic just about every authoritative medical institution proclaimed that masks were not effective against the spread of a viral illness. Dr. Anthony Fauci himself stated during a *60-Minutes* interview in March 2020: "People should not be walking around with masks. It's not providing the perfect protection that people think it is. There are unintended consequences as people keep fiddling with their mask and keep touching their face which may actually increase the risk."

Others followed his lead. The CDC Director at the time, Dr. Robert Redfield, stated: "There is no role

for the masks in the community."
Even the World Health Organizaa-
tion (WHO) weighed in: "The use of
masks is insufficient to provide the
adequate level of protection and other equally relevant measures should
be adopted." *The New England Journal
of Medicine* published a very long article from doctors at Harvard Medical School and Massachusetts General Hospital.

The article noted: "We know that
wearing a mask outside healthcare
facilities offers little, if any, protection
from infection." They went on to say:
"It is also clear that masks serve sym-

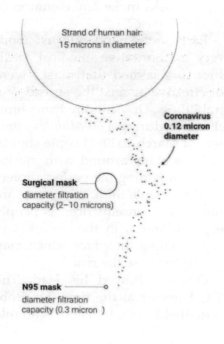

Strand of human hair:
15 microns in diameter

Coronavirus
0.12 micron
diameter

Surgical mask
diameter filtration
capacity (2–10 microns)

N95 mask
diameter filtration
capacity (0.3 micron)

bolic roles. Masks are not only tools, they are also talismans that may help increase health care workers' perceived sense of safety, well-being…" Finally, the NEJM article concluded: "Expanded masking protocols greatest contribution may be to reduce the transmission of *anxiety*." [Emphasis added.] The *Journal of the American Medical Association*, not wanting to be left out, reported: "…there is no evidence to suggest that face masks worn by healthy individuals are effective in preventing people from becoming ill."

These are just a few of the conclusions of authoritative health experts that advised against masking the healthy population. But then politics got in the way of scientific evidence.

Consider just this: The virus particles associated with Covid-19 are about 0.12 microns in size. Why does size matter? Because the pores in a typical surgical mask can only filter objects that are 3.0 micons and larger. An N95 mask, if fitted properly to the individual wearer, can filter objects down to 0.1 to 0.3 microns. But a 0.12 micron Covid-19 virus particle will pass through an ordinary surgical mask easily — like the light blue paper masks worn by the majority of Americans — as well as any homemade cloth product.

These masks are the equivalent of erecting a chain-link fence to keep mosquitoes out.

Interestingly, there have been several studies looking to prevent the transmission of influenza and the common cold using surgical masks. The studies concluded they were ineffective. That's why we are not told to wear them during flu season. There are no formal studies yet related to masks and Covid-19 — only anecdotal evidence which is mixed. Analytical studies have shown many of the counties and states with the strictest mask policies fared no better than lenient counties and states and some even worse.

Fair enough, but a lot of readers will have heard that the Covid-19 virus is mostly carried in droplets of mucus or spittle and any face covering will stop that. There is some truth to this. A sneeze or cough containing Covid-19 may be partially blocked by any of the commonly used masks or a bandana. The science is currently mixed on the extent to which a sneeze or cough moves through these protective coverings. But note this: As moisture builds up inside a face covering, its filtering ability drops precipitously.

Moreover, if Covid-19 virus particles are now trapped and building up inside a face covering, it could be re-infecting the mask wearer, going from the mouth into the nose. This does not even account for the fact that every time the mask is adjusted or touched the virus may be transferred to the wearer's fingers or hands and whatever they subsequently touch. Another point: Please never exercise in a mask - even a brisk walk in a mask can lower your oxygen saturation, cause headaches and put your health at risk. And, one last point: Remove your mask if you are driving alone in your car — you look foolish.

Why, then, does a surgeon wear a mask during an operation? It is *not* to prevent a viral infection impacting the patient. It is to prevent the surgeons' and nurses' spittle, while talking during surgery, from getting into an open wound and causing a nosocomial bacterial infection. It is also to prevent pieces of tissue splattering into the surgeons' and nurses' mouths. If there were no need to talk during surgery, there would be little need to wear a mask – other than splattering tissue. Furthermore, a 2008 study of 53 surgeons by A. Beder showed reduced oxygenation and increased pulse rate after an hour of mask-wearing during surgery.

What if I'm six feet away from another person, am I safe? There is zero science behind the six feet of separation mantra. The Europeans and Africans maintain a meter (about 3 feet) of separation. But why not 4 feet or 8 feet? Because six feet was thought to be an easily remembered and calibrated number. It is a guess that most people can't project the particles of a sneeze or cough further than six feet.

Common sense tells us that if you are not near an infected person you are less likely to get infected - which is why it made no sense when the California Governor closed outdoor spaces such as parks and beaches. Separation there among non-family members is normally more than six feet. But since when have politicians applied common sense?

Okay, back to size - the size of children! Little people are at very low risk to contract Covid-19. The CDC's web site reports the risk of death in children from Covid-19 as 0.0%. Children have a 50-times higher risk of drowning and a much higher risk of being killed in an auto accident than dying from Covid-19. In fact, masking children in school causes more harm than good. In children we see skyrocketing anxiety, depression, learning disorders and even suicidal ideation. These injuries will continue long after we lift the unscientific school mask mandates now in place.

We expect to see a Post-Traumatic-Stress-Disorder-like syndrome in children lasting long into the future. Fear is a powerful facilitator of public opinion. It is often used by organizations to galvanize action from their members. As to masks, the demands of many of the teachers' unions are driving the narrative to wear masks when schools are open. The concern, as expressed by the leadership of some teacher unions, is that a little one might be able to spread the virus to an at-risk teacher, despite evidence to the contrary.

I have a patient who has Stage 4 breast cancer and is undergoing chemotherapy. This depletes her immune system, making her at a high risk of getting an opportunistic infection such as Covid-19. We can either protect her — and if necessary isolate her — or we can mask everyone who lives near her or visits wherever she goes. Common sense tells us it is far more efficient and effective to protect *her* than worry about everyone around her. More importantly, we have multiple studies that show there is little evidence that children can asymptomatically transmit the virus. The WHO indicated as such in June 2020 when Maria Van Kerkhove, M.D., the technical lead for the WHO's pandemic response unit, said: "...asymptomatic transmission appears to be very rare." Yet, political pressure caused her to walk back her comments within a day. The politics of healthcare is getting out of hand and in the grasp of the wrong people.

So what should we do now? Mask up if you feel you must, but don't claim it is because of science. The political winds buoyed by science are now pushing state and the federal government to ease up on mask mandates. A school child, in my opinion, should never wear a mask and socially distancing among children also makes no sense at all.

A healthy society protects the most vulnerable and isolates those who are sick. Covid-19 poses only influenza-like risk to the young and the healthy. It is time we stop virtue signaling and actually follow the science.

The damage done by the government edicts of tyrannical leaders will be long lasting in both the harm done to our citizens and in our long-term memories. We should size up this pandemic properly, ignore the sensation-mongering headlines and act with a touch of old fashion common sense.

BETTER TO BE
GOOD OR LUCKY?

AS EXPECTED BY most observers, the number of cases of Covid-19 continues to fluctuate as the Delta variant or any successor arrives on the virus scene.

There are primarily three reasons for this:

1. More testing is being performed.
2. People are moving about as the economy opens up.
3. The media continues to do what it does best — use scare tactics to drive testing.

The net statistic that will be used to frighten the American people will be the increase in hospitalizations that will follow the expanded number of cases. Days or weeks after hospitalizations grow in number, even more deaths are likely to arise.

What will not be reported

in the gloomy news roundups is the fact that we now have effective treatments for Covid-19. More appalling, many doctors and hospital systems are refusing to use these effective therapeutics.

Dr. Vladimir Zelenko, a New York family medicine physician, has pioneered a treatment strategy that works well but is still shunned by most of the medical profession and ignored by the mainstream news media. Dr. Zelenko said this on Dennis Prager's radio show on July 10, 2020:

> I don't care what 'they' say anymore, I would rather speak directly to the American people and tell them I have some very good news for [them]. We have an answer to the terrible infection, we have a very effective way of treating it. In the high-risk groups there is a 99.3% survival [rate] and a 84% reduction in hospitalizations. There is also a 100% survival rate in low-risk pa-

> *tients when treatment is start-*
> *ed in the first five days [after*
> *the onset of] an infection.*

One of the key reasons for the increase in hospitalizations has also not been widely reported. The increase is in part due to the reopening of hospitals to elective surgeries and procedures. These had previously been categorized as non-essential by many state authorities. Hospitals are also using ICU wards to isolate Covid-19 cases and not for the usual ICU acuity care. In addition, some of the ICU admissions are for relatively routine, not Covid-19, hospitalizations.

The early treatment of Covid-19 patients with mild symptoms has proven overwhelmingly effective. In Texas, Dr. Richard Bartlett has a 100% track record of no deaths with his treatment of Covid-19. As part of his protocol, he uses an inexpensive inhaled asthma steroid called budesonide. This treats the pulmonary inflammation that is often the culprit in the death of patients with Covid-19.

Both hydroxycholoquine and budesonide are only a fraction of the cost of Remdesivir, with its multiple thousands of dollars per dosage. Is there a financial incentive in maligning the cheaper treatment alternatives? One has to suspect that.

Dr. Zelenko's protocol of hydroxychloroquine + zinc + azithromycin has been made publicly toxic because President Trump mentioned just the first ingredient during a briefing on March 19, 2020. Yet a study released on July 2, 2020 by the Henry Ford Health Systems in Detroit showed a significant reduction in deaths among more than 2,500 hospitalized patients using the Zelenko cocktail of medicines.

For reasons that are not clear, the national media has refused to acknowledge this hydroxychloroquine study result, perhaps for fear that President

Trump might be given some credit for promoting the drug. Information on the study can be found on the Internet under "Henry Ford Health study."

Chloroquine — a derivative of quinine, the extract of the bark of the cinchona tree — was developed in the 1930's as an alternative for the treatment and prevention of malaria. As most are aware, malaria is a mosquito-borne illness that has long been the scourge of the world's tropical regions.

Chloroquine was first used in a significant way during World War II in the Pacific Theatre. Later, soldiers in Vietnam were given weekly doses of chloroquine to prevent them from contracting malaria. All evidence points to its relative success in this usage with only 46 deaths out of more than 50,000 attributed cases.

Hydroxycloroquine is a better tolerated and more effective derivative of chloroquine, first approved for use in the United

States in 1955. It has been part of the effort of medical science to find a more efective way to deal with malaria.

The generic version of hydroxycloroquine has been around for many years and costs only pennies per dose. Studies eventually demonstrated that the drug also showed promise for the treatment of other chronic illnesses like rheumatoid arthritis and systemic lupus. By now, its safety is unquestioned and it has even been approved by the FDA for use during pregnancy and breastfeeding.

So, if you are someone who has recently tested positive for the Covid-19 virus, there is great hope for you. The challenge is to find a doctor who is not overly wary of the political waves generated by the disease.

Among the newest medications is Regneron's REGN-COV2. This was one of the medications used to treat President Trump's Covid infection. REGN-COV2 is a combination of two monoclonal antibodies designed to bind to one of Covid-19's distinguishing spikes. This reduces the viral load and the clinical symptoms.

Compare REGN-COV2 to Remdesivir. It works by imitating a building block of the virus's RNA. Just as a defective Lego block prevents the expansion of a Lego structure, so a defective element of the virus's RNA stops it working properly. Add zinc. It disrupts the virus's replication mechanism to create a potent trifecta of medicines to help return those infected by Covid-19 to good health.

If you are not currently positive for Covid-19, there is a lot you can do to maximize your ability to block the infection from coming your way.

First and most important, take care of yourself by eating healthful food, exercising daily, maintaining your proper weight and taking an immune-enhancing supplement.

Check with your own physician prior to using any medication. Here is a list of the supplements I recommend to my patients to keep them from getting infected:

- Zinc — 25 mg daily
- Vitamin C — 3,000 mg daily
- Vitamin D3 — 5,000 IU daily
- Melatonin — 1-2 mg at bedtime
- Quercetin — 500 mg daily
- Fish oil — 3,000mg of EPA + DHA

In addition, stay well hydrated, avoid sugar and excessive alcohol. Get plenty of fresh air and sunshine, sleep six to eight hours each night, and most important, manage your stress.

Should you contract Covid, work with a doctor who understands that effective treatments are available outside of a hospital setting.

NOT DYING IS BAD FOR HEADLINES — COVID-19 CASES VS. DEATH

AS WE TRAVEL the bumpy Covid-19 path back to normalcy, you would never know that we are headed in the right direction.

Teasers on television and headlines in newspapers report a "surge" in new cases of Covid-19 as well as hospitalizations; the

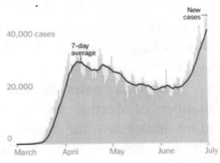

New reported cases by day in the United States

40,000 cases

7-day average

New cases

20,000

0

March   April   May   June   July

41

the teasers and headlines, of course, fail to mention that the most important measure of our progress in getting ahead of the virus is the number of *deaths* recorded. The fact that the rate of fatalities to infection is very low now is seldom if ever mentioned.

After weeks of unnecessarily shutting down our economy and the travesty of closing schools, it was certain that cases of Covid-19 would spike as we reopened society — *duh*.

When the massive protests and riots around the county over the succeeding weeks are taken into consideration, it ought to be no surprise that cases are increasing. It is important to remember that the purpose of "flattening the curve" was to delay cases and deaths from Covid-19 to a future date when our healthcare capacity was no longer threatened with being overwhelmed.

The great news is that the number of deaths, and perhaps more important the fatality rate (the likelihood of death from infection), have plummeted. Why? Because this virus tends to be very mild in the young and healthy — the cohort that has been most susceptible to infection of late.

The CDC data show that as of July 1, 2020, deaths among people younger than 25 are fewer than 170 out of a total of 120,000. It is almost non-pronounceable to state a fatality rate — too many zeros after the decimal point.

Covid-19 Deaths By Week They Occurred

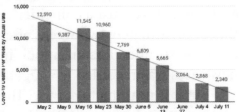

My point is this: If you are young and healthy, you have nothing to fear from the coronavirus except fear-mongering from the media.

The average age of the "new cases" now spiking is 31 years old — similar to the age, by most estimates, of the protesters and rioters recently in the streets. I looked carefully at news coverage and videos of the protests and riots across the country and couldn't find many participants who looked 65 years or older.

Moreover, I couldn't see much social distancing going on either! Yet, we are told that reopening restaurants, bars and retail stores — most of which had tried to arrange for social distancing and some of which required masks — is the source of the spike in new cases. I guess it was just a coincidence that approximately two weeks after the protests and riots an uptick in new cases was recorded.

But it is nothing to die for: The reason for a drop in deaths from Covid-19 is because a younger group is being exposed to the virus and we are much better at treating it.

Hydroxychloroquine + Z- pack + zinc have proven effective for early and mild infections. Doctors are also now using anti-inflammatory steroids (Dexamethasone and Prednisone) earlier and more aggressively in treatments.

The length of hospital stays for Covid-19 is also now dropping. Earlier I wrote that the Covid-19 fatality rate will turn out to be in the ballpark of a bad influenza season. This is now becoming the case. The more we test, the lower the fatality rate as well. Approximately 98.6% of infected people will experience either no symptoms or mild symptoms and approximately 99.85% of all infected cases will recover.

Here are relevant CDC statistics on survivability of Covid-19 by age:

AGE RANGE	RATE
0-19	99.997%
20-49	99.980%
50-69	99.500%
70+	94.600%

A healthy society protects its most vulnerable members and isolates the sick while allowing the healthy individuals to go about their lives. Let's celebrate the drop in deaths and stop sensationalizing the "new cases" of Covid-19 as worthy of another shutdown.

SHE DIED ALONE

SHE WAS ALONE in her room in the nursing home. For the past two months she had been receiving hospice care for heart failure and a chronic lung disease. Several years before, her family had made the decision to transfer her to the nursing home as her health care needs became unmanageable for the family.

As a mother, grandmother and sister, she had plenty of visitors — often several times a week and always on Sunday. She seemed happy. The staff at the nursing home was very attentive and the facility itself was clean, comfortable, even homey.

Then Covid-19 hit in March 2020 and everything changed for this family: No visitations were allowed, no meals could be shared, no hugs or kisses permitted, no one to get closer than two arms' lengths away. The escalation from resident to hospice care occurred soon after the pandemic began lurking around the facility. The family believed her deterioration could be correlated to the withdrawal of human touch from her loved ones.

She passed away in May with no family at her bedside, no one to hold her hand, no friend or relative to whisper something lovingly in her ear. While she died peacefully, she left this world utterly alone. It didn't have to be this way.

The circumstances of this story have been repeated thousands of times across the country and the world. Almost 50% of all Covid-19 deaths in this country have occurred in nursing homes. Most deaths in nursing homes, however, are not *due* to Covid-19.

There are countless other causes of death. Yet all visitations have been banned for everyone in a U.S. nursing home. How does it make any sense to condemn an entire population of every nursing as safely? Surely being able to pray is home in America to loneliness? Could such an edict do more harm than good?

Isolation is deadly — we have seen the highest suicide rates since the Great Depression. More children have died from suicide than Covid-19. The shutdown of houses of worship, schools and businesses has had devastating consequences. It may take us years to recover from this foolish decision.

Consider this:

- If we can fly crosscounty safely with masks on, why can't we be with a loved one during their final days utilizing the same precautions? We have rapid Covid-19 testing that provide results within an hour. Couldn't these be employed by nursing homes to ensure that visitors are negative for the coronavirus?
- If we can shop safely at Costco with masks on, why can't we worship at a church, synagogue or mosque just

just as important as buying merchandise at a discount.

- If we can wait in line to buy a bottle of liquor or a six-pack of beer without danger, why can't we just as safely wait in line to vote on election day? Wouldn't all of us consider voting equally important to getting a buzz on?

- Why can't children attend school in person, but adults can attend a political rally if they like? Children are at very low risk for contracting Covid-19. Note that as of August 18, 2020, only one child under the age of 18 has died of Covid-19 in the state of California, whereas many more have died of influenza. We do not routinely close schools due to an outbreak of influenza, yet we are closing them all over the country because of the threat of this virus. Our children are unnecessarily suffering from being out of school. I wish the fact that child abuse reporting is down 25% were the result of a less violent world. Unfortunately, we aren't that lucky. We have more abuse but less reporting.

Looking at these instances, the conclusion is clear: Covid-19 is no longer a healthcare crisis; it is a crisis of the soul of America.

I hope you will join me in pointing out these realities and inconsistencies to our decision-makers. There is hope that a growing number of Americans are waking up to the reality that our American way of life is in jeopardy.

Like generations before us who waged war to protect America, we are being called upon now to fight for the country we love. Although we are not at war, our fight is equally important and difficult.

Fight by questioning everything. Fight by supporting candidates for local office (school board and city council) who understand what is at stake. Fight by running for office yourself. Fight by peacefully protesting. Fight by writing letters to the editor or posting on a social media site. Fight by being a little uncomfortable in speaking out. I am.

I cried writing this piece while thinking about the lady who died alone with no family members nearby. It is so very sad for those that are unnecessarily suffering. Join me in fighting to preserve liberty in our land.

Notes/Questions

BOOSTERS

THIS ESSAY IS PRESENTED IN A SMALLER TYPEFACE
BECAUSE OF SPACE LIMITATIONS.

PFIZER AND OTHER MAJOR PHARMA houses are now pushing the idea of giving an extra dose of vaccine to those who have already had their original shots.

The media is calling these added injections "booster" shots — but in reality they are an exact duplicate of what the first and second shots contained. In my view, it is a repetition of the inaccuracy of the claims made when the vaccines were originally introduced.

The data from the companies show that the efficacy of their Covid-19 vaccine drops precipitously after 6 months. In Israel, one of the most highly vaccinated countries in the world, the Ministry of Health has reported that the Pfizer vaccine is only 39% effective against the Delta variant of Covid-19.

Here is a problem with the current conversation between the experts and the media:

If the original vaccines were so good — as most "experts" seem to agree — then why do we suddenly need "boosters"? Moreover, if the vaccines were also so good at preventing the disease from taking hold in a person's body, why do vaccinated people have to wear masks or be wary of unvaccinated people?

Meanwhile, more data are now indicating that natural immunity from having had Covid is far superior to the immunity provided by any of the vaccines. If that is true, why then is the government pushing to vaccinate Covid-recovered patients along with the unvaccinated?

The questions about the vaccines keep piling up and the answers seem as slippery as the mud coursing down a hillside in a heavy rainstorm. Could it be that financial incentives are behind the push for more vaccines into more arms? If "dollars" are what the "booster" shots are all about, who will provide the "sense" to the American people?

Recently, the Food and Drug Administration's vaccine advisory panel voted overwhelmingly *against* recommending approval of a booster shot of Pfizer-BioNTech's vaccine for people 16-years and older. It said the data in hand did not justify the government's request to give boosters to everyone. The panel did, however, endorse an additional dose for those most vulnerable to the virus — the immune compromised and those above the age of 65.

Pfizer has said that the Covid-19 vaccine efficacy weakens over time and hence the need to give the immune system a "boost." But recall that it wasn't long ago that we were told that the vaccines were at least 95% effective in their clinical trials. It was encouraging news.

The real world, however, is vastly different from he environment surrounding carefully controlled tests. A Kaiser Permanente study suggest protection against Covid-19 infection dropped from 88 percent in the first month after receiving the second dose to just 47% after five months. That is a significant difference.

Moreover, two senior FDA officials and a dozen top researchers argued that the booster shots aren't needed for the general population. They said that the potential side-effects from extra doses could outweigh any benefits to the vast majority of citizens. They, at least, remembered that element of the Hippocratic oath mandating *primum non nocerre* — first do no harm.

The World Health Organization (WHO) has also weighed in on the booster issue. Director-General Tedors Adhanom Ghebreyesus of Ethiopia said: "I will not stay silent when companies and countries that control the global supply of vaccines think the world's poor should be satisfied with leftovers."

Ghebreyesus didn't mention another problem which might alleviate the Third World vaccine shortage. The United States long ago promised to share the vaccine technology so that other companies could begin manufacturing it for the millions who lack it. But the U.S. seems to be dragging its feet on releasing the necessary information. No sharing has occurred. Could it be that this is also more a matter of dollars than sense?

Meanwhile, Britain's vaccine advisory body has recommended against universal Covid-19 vaccinations for healthy teens between the ages of 12 and 15. The Brits believe that any marginal benefit is not worth the risk. Not surprisingly, this was despite a government push to proceed with a plan to jab as many young arms as it could reach. As a reminder, the CDC's own data show that the survivability of Covid-19 in kids less than 18 years old is 99.997 percent. How many young lives does that actually leave at risk?

To further complicate the vaccine and booster narrative, the state of Massachusetts reported over 4,500 new breakthrough cases for fully vaccinated people during one week in late September 2021. A breakthrough case occurs when someone gets

Covid despite being fully vaccinated. We are seeing this across the county.

I have also been witness to it in my own practice — multiple cases of fully-vaccinated patients getting Covid. Talk radio host Mark Larsen, who mentioned on air that he is a patient of mine, had me on his progam to discuss his situation. I noted that he had received Regeneron and that he is now recovering from a post-vax covid infection.

While hospitalizations from Covid are more likely in the unvaccinated, this may change as more and more people get the shot. In Israel, approximately 75 percent of all Covid hospitalizations are now for fully vaccinated patients!

I should also point out that US hospital data on Covid is very misleading. That data include a group that are in the hospital for medical issues *unrelated* to Covid, but eventually test positive for the disease.

An article in *The Atlantic* recently pointed out that "almost half of those hospitalized with

Covid-19 have mild or asymptomatic cases." The article also notes that "...there are many Covid patients in the hospital with fairly mild symptoms [but] who have been admitted for further observation on account of their comorbidities, or because they reported feeling short of breath."

Finally, *The Atlantic* reports that "... roughly half of all the hospitalized patients showing up on the COVID-data dashboards in 2021 may have been admitted [to a hospital] for another reason entirely, or had only a mild presentation of [the] disease."

So what to make of all this? Here is my perspective:

The various Covid-19 vaccines are not as advertised. That is, they help around the margins in preventing severe symptoms from the disease but do not prevent infection or the spreading of the virus. They are failing to deliver on what was promised.

To reinforce this point, we are seeing more and more vaccine failures and significant injury numbers coming into the Vaccine Adverse Event Reporting System (VAERS) — the specialized website run by the FDA and CDC.

In my view, we could never vaccinate our way out of this pandemic. Rather we needed to refocus our medical resources on early treatment of those ill and on protecting the vulnerable.

There is growing evidence of the efficacy of repurposed medications and monoclonal antibodies such as Regenerone. In addition, Merck has a new pill said to reduce hospitalizations and death by half in those who have Covid.

It is clear. The public must have easy access to these medications and physicians must be advised on their use. If not, patients will continue to suffer and the vaccine push will continue to eat up valuable resources.

PUT YOUR
DAMN MASK ON

"PUT YOUR DAMN mask on," said
a guy three seats behind me. We
were both on a Southwest Air-
line flight to Washington, DC.
The loudmouth was talking to a
passenger in the aisle. His mask
had slid down under his nose as
he negotiated his carry-on bag
while holding his little daughter's hand.

I turned around to see who was making such a rude demand. Just as I suspected, it was the guy who came aboard sporting a full N95 mask, face shield and surgical gloves. Forget for a moment that the father actually posed zero risk to the hazmat-protected passenger. Besides, who appointed him a member of the mask police and protector of the well-being of all the other passengers? Where did decorum and common courtesy go?

When you cannot see someone's face, anonymity prevails. It provides the kind of disrespectful arrogance and brazen nastiness of the social media world. I remember hearing that you should not say anything about someone on social media that you would not say to them in person. The mask lets people say the nasty things that previously had been said while hiding behind a computer screen.

I am not arguing here for or against the efficacy of masks — I have done so elsewhere in this book. I am simply advocating treating others with decency when the facts are in doubt or disagreements may prevail. It was clear by the look he gave me when I turned to see who was being so officious, that he believed his words were not only justified but should be re-inforced. I did the opposite.

"You're fine," I remarked to the passenger whose nose had become exposed. I slid my own mask off as I remarked, "Don't let him bother you" so the guy in the aisle could both hear me and see my supportive expression. From my standpoint, if someone is so worried about contracting the virus while wearing full hazmat gear, perhaps he ought not to be flying on a crowded airplane.

To further my evaluation of this passenger's psychological pathology, I noted that a German Sheppard was on the floor under his legs. Clearly, an "emotional support animal" for someone with doubtful stability.

What if another passenger were allergic to dogs? Should our hazmat friend be asked to take another flight or sit at the back? A psychiatrist friend, Mark McDonald, M.D., coined a phrase that fit this guy perfectly: "The hyper-evaluation of self leads to despicable behavior."

I do not doubt that hazmat guy thinks highly of his righteous behavior and his enhanced self-esteem. Fear is a powerful accelerant of psychological pathology. Perhaps I should have shown hazmat guy what was written on the napkin accompanying my cup of ice water: "In a world full on no, we're a plane full of yes." Except that there is no "yes" to personal liberty.

I don't know where this is leading but utopia is not the

next stop on this road — quite the opposite. The train to hell is being conducted by the likes of hazmat guy, Governors Newsom (California) and Whitmer (Michigan) and former governor Cuomo (New York).

I find myself only wanting to be with those who are like minded. It seems no longer possible to disagree without being called names, to voice a contrary opinion without someone calling for your public shaming, or believe that liberty has a higher value than the status quo of one's health.

I worry for the next generation who will grow up without insight into America's past and without role models who believe in the founding principles of America:

- *E. Pluribus Unum*
- Liberty and
- In God We Trust.

For now I will continue to fight for America, be a role model of courage to my kids, and pray there are more of us than the hazmat guys to make a difference in the future of our country.

HERE WE GO AGAIN —
COVID-19/3.0

THIS ESSAY IS PRESENTED IN A SMALLER TYPEFACE
BECAUSE OF SPACE LIMITATIONS.

IT'S NOT ABOUT THE VARIANT, IT'S ABOUT THE FEAR. Like a magician who gains his reputation for sleight-of-hand — getting the audience to concentrate on one aspect of a trick while manipulating another — so government authorities around the world are using the unknown aspects of a new Covid-19 variant to keep the public in fear about their future well-being.

This chronic fear under the guise of an "emergency" has granted enormous power to those who the media anoint as "experts" and "scientists." They are milking their new-found fame and perceived wisdom to make pronouncements and exercise control without discussion, debate or appeal.

The South African medical association chairwoman, Angelique Coetzee, has stated that the Omicron variant has minor effects: "It presents mild disease with symptoms being sore muscles and tiredness for a day or two . . . those infected do not suffer loss of taste or smell. Those infected are being treated at home." Sounds pretty terrifying, doesn't it?

What is becoming clear, though, is that the current Covid-19 vaccinations are not fully effective against Omicron — or, for that matter, the Delta variant.

We are seeing more and more "breakthrough" cases — the fully vaccinated getting Covid and even dying of the infection. On the other hand, natural immunity seems to protect against all Covid-19 variants.

Booster shots are now being pushed as the solution to each new variant. Dr. Anthony Fauci even announced that new formulations of the current vaccines are not necessary, just get more boosters into arms. He made this statement despite the lack of science supporting his conclusion.

In some countries you are not considered fully vaccinated unless you have received the booster shot. There is no end in sight to what the "experts" decree. But worse, Fauci continues to ignore mounting evidence that the vaccines cause side effects and injuries to many people.

In short, the arrival of the Omicron variant has closed down travel again and created increased fear in our daily lives. It is unnecessary, harmful and counterproductive.

On a broader plain, Nobel economist F. A. Hayek and author of *The Road to Serfdom* got it exactly right when he wrote some years ago:

> *Emergencies have always been the pretext on which the safeguards of individual liberty have been eroded -- and once they are suspended it is not*

difficult for anyone who has assumed such emergency powers to see ... that the emergency will persist. [Emphasis added.]

Nazi propagandist Joseph Goebbels famously said:

If you tell a lie big enough and keep repeating it, people will eventually come to believe it. The lie can be maintained only for such time as the State can shield the people from the political, economic and/or military consequences of the lie. It thus becomes vitally important for the State to use all of its powers to repress dissent, for the truth is the mortal enemy of the lie, and thus by extension, the truth is the greatest enemy of the State.

One of the big lies of recent days involves the government's insistence on a single path to deal with Covid-19. There are much more effective ways than simply getting a vaccine, using a mask and remembering to keep one's distance from others. The government's one-size-fits-all solution has eroded our freedoms and put us on a collision course with totalitarianism.

Specifically, the federal government and most of the media claim, *falsely*, that the virus is equally dangerous to all, that we must be in mortal fear of it, that each of us, including our children, need to be vaccinated in order to work our way out of the danger, and that only government action, taken under emergency conditions, can save us from destroying our health care facilities and killing untold thousands of our citizens.

If anyone criticizes any of the officials of the CDC, the FDA or the National Institutes of Health, he or she criticizes science! Listen to Dr. Fauci on the point:

> *Attacks on me, quite frankly, are attacks on science. All of the things I have spoken about, consistently, from the very beginning, have been fundamentally based on science. Sometimes those things were inconvenient truths for people.*

That implies that science is always perfect, never subject to mistakes of any kind and always free of any possibility that data might be misinterpreted or erroneously gathered.

In truth, what is actually far more than inconvenient to the discussion of Covid-19, is how detrimental Dr. Fauci's statements have been. He has lied to the American people, stoking unnecessary fear among them. For example, medical professionals have increasing evidence of the successes of early treatment with repurposed medications and nutraceuticals to fight Covid.

Yet, Fauci has remained totally silent on the topic of early treatment. Why? Because the law clearly says that an Emergency Use Authorization for a new drug or vaccine cannot be granted if there are existing approved treatments available for use.

Well, many of us have been saying for two years that there are such treatments for Covid-19. But if Fauci and the other government doctors admitted this truth, the Emergency Use Authorization for the Pfizer, Moderna, and Johnson & Johnson vaccinations could not have been granted.

Moreover, we have not seen a single early treatment protocol endorsed by any of our academic institutions. Why? These same institutions rely heavily on government grants to continue their work; those grants are controlled by Fauci's agency at the NIH. Who would want to incur his wrath for fear of losing a future grant.

As a result, no one in government authority or connected to it at the hip by dollars has said a word about the importance of vitamin D3, for example; no one has offered a thought on the benefits of early use of a nasal rinse such as povidone iodine. If this nasal/throat protocol were used it could reduce the symptoms and duration of the illness and even prevent hospitalizations.

Early treatment works and it is a

travesty that our so-called experts are silent on the best of them. We are now expecting new prescription antiviral medications to be approved soon; these will be welcomed.

However, these new very expensive products will likely only marginally add to the effectiveness of what we can already accomplish with current early treatment protocols.

Fear, in fact, is proving more contagious than C19/O and more useful for population control than the truth. It is certainly of benefit to elements of big pharma — a sector drowning in profits provided by big government and allied with individuals at major academic institutions.

Note this: The Great Barrington Declaration was created in October 2020 by such prominent scientists as Dr. Martin Kulldorff, professor of medicine at Harvard University — a biostatistician, and epidemiologist with expertise in detecting and monitoring infectious disease outbreaks and vaccine safety evaluations; Dr. Sunetra Gupta, professor at Oxford University — an epidemiologist with expertise in immunology, vaccine development and mathematical modeling of infectious diseases; and Dr. Jay Bhattacharya, professor at Stanford University Medical School — a physician, epidemiologist, health economist and public health expert focusing on infectious diseases and vulnerable populations.

Their Great Barrington Declaration recommends an approach to Covid-19 called *Focused Protection*. That is, we should protect those who are most vulnerable — the elderly and those with significant underlying comorbidities — allowing those at minimal risk to live their lives normally without lockdowns or mandates. This declaration has subsequently been signed by more than 850,000 (!) scientists, physicians, and health professionals, myself included.

But the Declaration itself has been ignored and those who endorsed it have been discounted as a fringe group (850,000?!) advocating a risky proposition. The Infectious Diseases Society of America released a statement calling the Barrington plan "inappropriate, irresponsible and ill-informed."

Yet the approach advocated by the Declaration is now being proven correct. Covid-19/O, with its mild symptoms, seems another reason for adopting the Barrington approach.

The mask issue is another big lie perpetrated by big government in the thrall of its "experts" and "scientists." The masking policy has caused our children increased anxiety, depression and learning disorders. We have seen the highest suicide rate in teens that we have ever seen before.

Every major healthcare institution correctly told us, at the beginning of the pandemic, that masks are ineffective in keeping people safe from the spread of Covid-19 including Fauci, the World Health Organization, the *New England Journal of Medicine* and even the CDC itself — until fear of political reprisals forced them to retract their statements.

C19/O is a continuation of the pandemic of fear we have endured since 2020. We get what we tolerate and it is now past time for mass peaceful civil disobedience. Let's show our independence by refusing to abide by mask and vaccine mandates and by refusing to listen to the propaganda.

We must resist the lie that ineffective vaccines are the path to herd immunity. Consensus is the enemy of science and truth. This "trust the experts" style of policy making has led to an unprecedented loss of liberty and life.

As Robert F. Kennedy, Jr. recently said: "We must love our freedom more than we fear a germ." Liberty once lost is almost impossible to regain. We have to change the trajectory of the current policy or our country may never recover its place as one of the freest places on earth. Unfortunately, I am not optimistic the tyranny will end soon.

IT'S JUST
A TEST

I HEARD THIS startling story from a patient the other day: "I got a phone call . . . telling me my Covid-19 test came back positive. But I never had the test done, I told the caller. I checked in and completed the paperwork but I left because the wait was too long. They never [actually] tested me."

Incredibly, the caller didn't believe my patient. "No," she insisted, "I am sorry . . . your test results were positive." Worse, I suspect that these bogus "results" were sent to the state data bank that reports "new cases" to the CDC. One more error in the Johns Hopkins numbers that most of the media rely on.

I have read numerous other accounts similar to this on social media. They never seemed credible until several of my own patients reported directly to me that this is exactly what happened to them. Then, a member of my office staff reported the same experience.

Concerns about the accuracy of the multitude of Covid-19 tests being used are commonplace and with good reason. But now, the issue of ghost tests calls into further question the validity of the whole testing process itself.

Why would Covid-19 results be fabricated? Is it a simple clerical error? Are financial incentives being given for increased testing productivity? Are politics at play — the more tests administered, the better the bureaucrats in charge look?

But here's the important thing: If we can't get the basics right and if the numbers are wrong, how can we expect the experts to make the best possible public policy decisions for the country? We can't, and we can end up doing things that are actually inimical to the well-being of the people of this country!

Another real life story: One of my patients owns a business in the city of Long Beach, California. An employee of his, with whom he had direct contact, tested positive for Covid-19. As a result, he decided to get himself tested. So, he drove to a Covid-19 testing center.

The Sofia rapid nasal test was used at this site. This testing system uses an immunofluorescence-based assay. To my patient's consternation, he, too, tested positive. All of a sudden, he faced a cascade of unpleasant but necessary decisions including the cancellation of a long anticipated golf trip to Bandon Dunes, Oregon.

The next day he took his wife to the same testing center; he also decided to re-test himself just to be sure. Naturally, both he and his wife tested negative for the virus this time.

So which of the tests is correct, he wanted to know? Both my patient and his spouse had no symptoms. We are taught in medical school to treat the patient, not the results of tests. So, no treatment was instituted. But I wanted to know whether my patient's positive test from the day before had been reported?

Many of my patients that test positive go back and retest — sometimes multiple times until they get a negative result that conforms with how they feel. They do this for a variety of reasons: Legitimate concern about being contagious; return to work policies of their employers; and lack of confidence in any single test, to name a few.

If a patient were to test positive three times over a couple weeks, is every one of those positive tests sent to a central data bank? Could this be in part why we have seen the number of cases escalate? More positive tests lead to more multiple testing of a single patient. What if I test positive while on a trip to a different state and then test positive when I come home? Are my results counted twice with each state registering each new case? If the basic data begins to be suspect, how can you trust the effectiveness of the public policy on which it is based?

A recent Journal of the American Medical Association research article in JAMA Pediatrics indicated that children could be spreaders of the virus as they have a higher viral load when tested.

"Our analyses suggest children younger than 5 years with mild to moderate COVID19 [sic] have high amounts of SARS-CoV-2 viral RNA in their nasopharynx compared with older children and adults."

More fake news? What the JAMA article failed to tell readers is that the Covid-19 test the researchers used was the new Abbott Laboratory test that is not FDA-approved to provide quantitative data. That test is approved for qualitative data only.

In other words, the Abbott test is meant to answer whether there is virus present, not to determine how much virus is present. Scroll down to the end of the JAMA article and note what you see: Harsh criticism from physicians and Ph.Ds who call the findings reported in the article misleading at best and fraudulent at worst.

The author of the report is

William Muller, M.D., Ph.D. He is on the staff of the Ann & Robert H. Lurie Children's Hospital of Chicago, a part of Northwestern University's Feinberg School of Medicine. He responded to the outpouring of criticism this way:

> *We very much appreciate the attention this paper has received and the comments left by different readers. Several readers have left comments on technical characteristics of the assay used to generate the reported data. While it is correct that the clinical application of the assay is for qualitative detection of SARS-CoV-2 RNA, the cycle threshold data reported in this study were gathered for research purposes.*

Fine. But what does all that mean? The mainstream media ran pieces about this article and never saw the criticism calling the conclusions patently false.

It should not be forgotten

that just a short time before this episode, two of the leading medical journals in the world, The New England Journal of Medicine and The Lancet, both within hours of each other, retracted separate research articles that were critical of hydroxychloroquine

It is no wonder that so many people have lost faith in our so-called healthcare experts and seem hopelessly confused as to what is to be believed about the dangers to their kids, their parents and themselves.

When the basics are wrong, how can anyone trust that the larger public policy decisions are right?

FEAR FATIGUE

OUR DAILY LIVES ARE FILLED with risks, lots of risks. Just look at these examples:

- Every year 30,000 to 40,000 Americans die in automobile accidents, yet none of us is willing to give up cars to avoid the possibility of dying in a car crash.
- Heart disease kills more than 600,000 of our fellow citizens annually, yet we continue to eat fast food and pack on extra pounds.
- Diabetes helps kill more than 80,000 Americans a year, yet sugar is still part of the food supply.
- Suicide takes the lives of close to 50,000 Americans annually, yet we have instituted Covid-19 policies that have increased the incidence of suicide

to the highest levels seen since the Great Depression.

- Millions of children are infected with influenza each year and hundreds die from the disease. But we have never closed our schools or insisted on masking the population to try to prevent the spread of flu.
- Child abuse and child sex trafficking are at record levels and many specialists believe that it is due, in part, because our schools are closed while adults are unable to go to their normal places of work.

The CDC estimated that as many as 500,000 people died worldwide from the H1N1 virus in 2009 — the first year that virus circulated. Overall, 80% of H1N1 virus-related deaths were thought to have occurred in people younger than 65 years of age. Despite this, we didn't close the schools, mask the population, or shut down the economy.

In 1968, the Hong Kong flu killed approximately four million people globally. Our economy remained open in the face of those staggering numbers; moreover, the three-day Woodstock festival in upstate New York rocked on during the epidemic.

Isn't it time we stopped living our lives and dictating what we can and cannot do based solely on fear of every new danger? Instead, shouldn't we let each individual decide what risks he or she is willing to take?

The famous declaration in President Franklin D. Roosevelt's first inaugural address seems as appropriate today as it was in 1933:

> *The only thing we have to fear is fear itself.*

We are Americans — the freest, most prosperous society in the history of the world. We fought to end slavery, losing over 600,000 people in the Civil War. We have fought in two world wars and played a dominant role in winning both. Those wars cost the lives of more than 500,000 Americans.

We take risks to enjoy the freedom they gave us — or at least we used to. We ride motorcycles, skydive, climb mountains, and drive too fast. We eat the wrong foods, drink too much alcohol, and live in ways that are often not healthful; but, we cherish our right to do so.

"Better Safe than Sorry" has NOT been elevated to a national motto. It doesn't appear on any flags; no one is rushing to buy T-shirts emblazoned with those words. But, "Live Free or Die"? That's a different story. It is the motto of the state of New Hampshire and was a colonial rallying cry. The words " . . . and the home of the brave" end the national anthem. Do we still believe it is?

Fear is far deadlier and more contagious than Covid-19. Fear raises our blood pressure to unhealthful levels; fear influences us to make poor decisions. We fear being criticized, we fear exposing our ideas, we fear offending others, and we fear being infected by a virus that is not much more deadly than viruses of the past.

As a physician with more than 25 years in practice, I will tell you what I tell my patients: You should not fear Covid-19. You should properly prepare and protect the most vulnerable in your homes, your businesses and in society; but you should join everyone else in living your life in maximum liberty with common sense protections and precautions.

We have an effective treatment when symptoms of

Covid-19 are seen early and remain mild. These treatments can also offer protection for the most vulnerable.

Stop listening to those that want you to stay in a state of chronic fear. Turn off the mainstream media constantly using fear to capture your attention. Don't let those who want power over our lives to gain more of a foothold in Washington.

Turn on friendships and optimism and life. Turn on church and community and hope. Live with purpose, and fear fatigue will never become a problem.

The Covid-19 Vaccines: Risks and Benefits

THIS ESSAY HAS BEEN UPDATED AND EXPANDED. IN ORDER TO ACCOMMODATE IT TO THE SPACE AVAILABLE, THE TYPE SIZE HAS BEEN REDUCED.

EVERYONE — QUITE LITERALLY — IS well aware of the vaccines being administered around the world to slow Covid-19. Even though the vaccines in use in the United States appear to be effective in reducing the spread of the virus, what is forgotten is that the Covid-19 has only recently been formally approved by the American government's Food and Drug Administration (FDA).

Before that, all US Covid-19 vaccinations were administered under the FDA's Emergency Use Authorization (EUA) rules. These rules allow healthcare agencies to distribute and administer vaccines to designated groups when there is no other approved countermeasure against a threat to public health.

It typically takes many years to create a new vaccine against an illness and even then success sometimes eludes the best efforts of the scientists working on the problem. For example, we have no vaccination yet against HIV, hepatitis-c, or the common cold — just three illnesses that have long caused misery to many.

It should be noted, however, that the World Health Organization has just endorsed a malaria vaccination — one that attacks the mosquito-borne parasite that claims a victim under the age of five every two minutes. Time will tell, of course, how effective this new vaccine for one type of malaria proves to be.

The vaccines used for the first Covid-19 inoculations, brought to market by Pfizer and Moderna, employ a messenger RNA (mRNA) technology. Significantly, this is the first time this particular mechanism has been used in a vaccine. For the most part, mRNA technology has been used in cancer therapy because of its success in producing various proteins to attack and disrupt the operation of certain cancer cells.

Most of the commentary from "health experts" has suggested that it was not too much of a leap to use this approach to develop vaccinations. Normally, the DNA in the nucleus of a human cell produces mRNA. The mRNA acts as an instruction manual to tell the cells how to create proteins.

The idea behind its use in an anti-Covid-19 vaccine was to produce a synthetic mRNA that instructed the cell's ribosome protein factory to create a SARS CoV-2 spike protein. The appearance of a spike protein then stimulates the body's own immune system to create an antibody against the spike. Technically, then, the use of mRNA in this way is not a true vaccine, but rather a type of immunologic or gene therapy.

The vaccine developed by AstraZeneca for Covid-19 uses a different platform. It takes a piece of genetic material from the SARS CoV-2 virus and inserts it into a common cold virus — in this case an adenovirus. The genetics of the adenovirus vector are then altered so it is not infectious. When the adenovirus enters a cell, it stimulates the creation of the spike protein.

The good news is that the AstraZeneca vaccine can be stored at normal refrigerator temperatures for up to six months. The disappointing news is it seems to be less effective than either Pfizer or Moderna. While AZ's vaccine may become the preferred one in less developed countries because of storage issues, it has shown significant side effects to the point that many European Union countries (including Germany, France, Italy, Spain, and most recently Denmark) have banned its use for fear that it may cause dangerous blood clots.

Johnson & Johnson's Covid-19 vaccine also uses an adenovirus transport mechanism to get the SARS CoV-2 DNA into the cells in order to create the desired antibody response to the spike protein. At one point the FDA and CDC put a pause on the use of the J&J vaccine in the United States because of a potentially dangerous blood clot reaction occurring a week or so after a person receives the shot.

Following an 11-day "pause" in the J&J vaccination program to review the situation, the FDA allowed use of this vaccine with an added warning to those who receive it.

Importantly, the U.S. regulatory bodies are acknowledging a causal relationship between the J&J vaccine and this rare but serious side effect. As of mid-April 2021, the J&J vaccine is not approved in the UK. The European Medicines Agency (EMA) decided that a warning about unusual blood clots would suffice on its labels, just as the health agency did for the AstraZeneca's Covid-19 vaccine administered in its area of responsibility.

One further note. The adenovirus used for both the Astra-Zenica and the J&J vaccines are grown in aborted fetal cell lines and the Pfizer and Moderna vaccines were tested using fetal cell lines. This has raised understandable concerns among many religious groups.

Finally, this: A new vaccine by Novavax may have an added advantage over the currently approved vaccines. It does not use mRNA technology to induce the body to create the SARS CoV-2 spike protein. Rather, the Novavax uses an altered spike protein to create an antibody response. As such, the amount of spike protein is limited and therefore may create fewer side effects and less potential long term complications.

Because of these caveats and side notes to all the vaccines, is it any wonder that 30% of US healthcare workers are indicating that they will pass on getting any Covid-19 vaccine. Forty percent of Marines in a recent poll are also very wary of getting vaccinated against a disease that most soldiers see as a low risk of having serious consequences compared to what else they face on a day-to-day basis.

Here are more areas of concern about the Covid-19 vaccines that the public needs to take into consideration:

- Could the vaccines in use now create asymptomatic carriers who transmit the virus to others?

- Are the vaccines protective against the new SARS CoV-2 variants — from the UK, South Africa, Brazil and India? An Israeli study has shown the Pfizer vaccine may put patients at higher risk for Covid-19 variants. The CDC now reports more than 7,000 fully vaccinated people have subsequently contracted Covid-19 and over 500 of them required hospitalization. We have now vaccinated more than 200,000,000 people in the U.S. Information about the various variant strains of SARS CoV-2 are still emerging. Some variants may prove dangerous, others not so much.
- Does the vaccine put patients at higher risk if the vaccine-induced immunity fails? What are the chances of long-term consequences that have yet to be seen?
- Most other vaccine development programs have first used lengthy animal studies to better assure

safety in humans. This was not done with any of the new Covid-19 vaccines. Are the clotting problems and other side effects now being studied the consequence of limited animal safety testing?

- Is it possible for the synthetic SARS CoV-2 mRNA to be integrated into the human host's cell genome? It has been seen before. Scientists at Harvard and the Massachusetts Institute of Technology produced findings about wild coronaviruses that raise questions about how viral RNA operates. This DNA-to-mRNA pathway is not always a one-way street. An enzyme called reverse transcriptase can convert RNA back into DNA allowing the latter to be integrated into the DNA in the cell nucleus.

- The high efficacy of a controlled, scientific study may prove to be very different from the real-world experience when massive numbers of people are involved. For example, data from the Centers for Disease Control and Prevention show that the efficacy of the influenza vaccination for the 2017-18 season was approximately 38 percent; only 20 percent was achieved in the 2018-19 season; and 39 percent for the 2019-20 season. When the influenza vaccination was first introduced in 1938, the efficacy was expected to be much higher than the current numbers.

- The CDC data show that the survival rate of those contracting Covid-19 goes up as age goes down. If you are less than 70 years old — the survival rate is 99.5 percent; if you are less than 50 years old — the rate jumps to 99.98 percent; and if you are under 20 years old, the chance of surviving Covid-19 is 99.997 percent! Why then would we consider vaccinating children without long term safety data? The risk of getting a

Covid vaccine seem to outweigh the benefits for kids.

- In fact, parents should know that seasonal influenza is a greater risk to the very young than Covid-19. Weighing the benefits versus the risks of accepting the Covid-19 vaccine could be a very difficult choice, especially for the young. It is unlikely, therefore, that I will recommend vaccinations for my young patients to protect them against catching a virus that more than 99 percent of them would survive without any effect.

- The same companies (and their executives) that look to profit from the new Covid-19 vaccines are immune from all liability should anything go wrong. In 1986, Congress passed the National Childhood Vaccine Injury Act (NCVIA). It provides immunity from liability to all vaccine manufacturing companies. Crony capitalism at its best — mandating purchase of a product by the public but no liability if it turns out badly. Rather than blanket immunity, why not establish national limits to damages along the lines of California's Medical Injury Compensation Reform Act (MICRA). And certainly no vaccines — or medicines for that matter — should ever be mandated by government.

- Minorities tend to be skeptical of the Federal government and especially of vaccinations administered by the U.S. Public Health Service. Why? The USPHS and the CDC carried out 40 years of secret syphilis experiments, using black populations as test subjects. Can we overcome this history to get the Covid-19 vaccine to minority communities? By June 2021, less than 10% of the black population of Florida had been vaccinated against Covid-19.
- How do we know that these

new vaccines will not cause significant problems from unforeseen side effects? We clearly don't.

- The CEO of Pfizer is now suggesting a third shot may be required to maintain protection from Covid-19 and then annual jabs thereafter. Have we really studied these vaccines thoroughly enough to stop guessing and make judgements based on scientific data?

- In 1976 we attempted a mass vaccination program against the swine flu with a then newly created vaccine. The vaccination program was aborted after 40 million doses had been administered. Over 500 people had come down with Guillain-Barré syndrome, a rare neurological disorder and there were 25 deaths associated with the swine flu vaccination. Will the memory of this epic government failure affect the further rollout of a new vaccine?

- We are now hearing rumblings that the Covid-19 vaccination could be made mandatory for air travel, international border crossings or even entering a theme park or government building. What could possibly go wrong trying to enforce that requirement?

- The vaccines are now being tested on children and pregnant women. This could be potentially dangerous as we simply do not know what the long-term effects of mRNA vaccines or the artificial creation of Covid-19 spike protein antibodies will be on children or fetuses. The CDC is now investigating dozens of cases of myocarditis (inflammation of the heart muscle) in young males following mRNA Covid vaccinations.

- Could a phenomenon known as antibody-dependent enhancement (ADE) occur with the Covid-19 vaccination. This reaction occurs when someone, exposed to SARS CoV-2 following vaccination, experiences a more severe reaction than if the vaccine were never given.

- A recent study published in JAMA showed SARS CoV-2 antibodies in breast milk following the onset of the Covid-19 vaccination program in breastfeeding women. What could the long-term implications of this be for the infants?

Does the risk/benefit ratio favor most people on the positive side of this equation? The website for VAERS (Vaccine Adverse Event Reporting System), which is run by the CDC and the FDA, is showing an alarming rise in reported side effects including over 5,000 deaths and thousands of ER visits associated with the Covid-19 vaccinations. There are more reported deaths with the Covid-19 vaccination than ALL the other vaccines in use over the last 15 years. The VAERS website requires doctors or patients to voluntarily report side effects — a passive reporting system. It is estimated that only 1 percent to 10 percent of all the adverse events are reported to this site as most people and even many physicians are unaware of the website's existence.

I am not opposed to vaccinations, per se, including the Covid-19 vaccination. But I am opposed to any order that infringes on a citizen's freedom of choice. So any government or private sector requirement for the vaccination would be wrong. By the same token, a vaccine passport as a prerequisite for travel or the use of any facility such as a hotel or restaurant should be opposed. I strongly support informed consent before any vaccination is administered and I oppose government mandated vaccinations for any purpose. Note that this is not the current U.S. standard.

While these questions are being answered and the other caveats mentioned above are resolved, we have to note that we are getting better and better at treating Covid-19: The death rate in terms of the total population continues to fall, hospital stays for Covid-19 are getting shorter, and hospital mortality from Covid-19 has dropped significantly. Early multi-drug outpatient treatment with repurposed medications is showing great success. There are a greater number of physicians embracing this approach and even the press and social media have finally allowed this message to get through their editorial gauntlets. Had we been allowed to advocate for early treatment with repurposed medications earlier in this pandemic, I believe we would have seen significantly less deaths from Covid-19.

In conclusion, a Covid-19 vaccine should be viewed as one of many tools to stop the spread of the pandemic but not as the magic bullet to end it. Skepticism is a good thing for science and extending that attitude toward these vaccines is no exception.

Concluding Thoughts on Covid-19

THIS ESSAY IS PRESENTED IN A SMALLER TYPEFACE
BECAUSE OF SPACE LIMITATIONS.

AS THE PANDEMIC'S GRIP ON the actual and potential health of so many Americans lessens and as parts of the world opens up to a return to normalcy, we are gaining some clarity on a number of the issues that have arisen over the past two years.

While final conclusions about Covid-19 are still not warranted, here are my current thoughts on how we have dealt with this pandemic:

- Our reaction to Covid-19 has caused more harm than the virus itself.
- The decisions made about the pandemic were never based wholly on science, they were always partially based on politics.
- Cancel-culture — the censoring of medical opinions that are not in-line with the health care establishment — has unnecessarily harmed hundreds of thousands of people.
- The origin of the SARS CoV-2 virus seems more likely to have come from the Wuhan virology lab through dangerous gain-of-function research funded in part by US tax dollars, not from wild animals.
- The Covid-19 *vaccines* in young people cause more harm than good. Children should never be vaccinated against Covid-19 where there is little risk.
- Covid-19 is an extremely easy infection to overcome for the young and healthy or if you well-established and inexpensive treatment is started immediately upon the onset of symptoms.
- The refusal of pharmacies to dispense re-purposed medications for Covid is wrong and has directly led to unnecessary delays in treatment and in many cases unnecessary hospitalizations.
- Studies continue to show that early treatment with repurposed medication such as Ivermectin and Fluvoxamine work

welll when used early. Moreover, the importance of Vitamin D and other supplements in the prevention and treatment of Covid-19 continue to be underreported.

- The *natural* immunity of those recovered from Covid-19 is stronger and broader based than immunity provided by any of the vaccines in use. A Covid-19-recovered patient does not need to be vaccinated and may be at higher risk of significant side effects if subjected to a vaccine.
- Masks have never worked to stem the spread of a virus from person to person and won't work in the future no matter how many "experts" argue to the contrary. Masks are only useful against a virus in order to create *fear and compliance*.
- Dr. Anthony Fauci, despite his new title as Chief Medical Advisor to the President of the United States, still has a lot to explain with regard to his involvement with gain-of-function research in Wuhan, suppression of early treatment modalities favored by many frontline doctors, misleading the public on the actual science behind masks, and pushing vaccines to be authorized under emergency-use-authorizations rather than proceeding under regular order protocols.

The bottom line, for me, is that in the future we must not allow the media and healthcare establishment to censor dissenting opinions. We must maintain a level of skepticism and curiosity in sifting through the opinions, studies and data. We need to always ask who benefits from each decision and most importantly what are the consequences of these decisions.

As Thomas Jefferson once stated: "Eternal vigilance is the price of liberty." Our freedoms have suffered under the guise of health and these freedoms once lost may never be regained.

ABOUT THE AUTHOR

DR. JEFFREY
I. BARKE is a
board-certi-
fied primary
care physician
who has a con-
cierge practice
in Corona del
Mar, California.
He completed his medical
school training and family practice residency at the University of
California Irvine and earned his
undergraduate degree at the University of Southern California.

During his more than 20
years as a physician, he has served
as an associate clinical professor
at the University of California Irvine's Medical School, chairman of
the Family Medicine Department
at Hoag Memorial Hospital, on
the board of directors of the Orange County Medical Association
and medical director of Pathways
to Independence.

Dr. Barke was elected to three
four-year terms as a member of

the Los Alamitos Unified School District Board of Education and is currently an elected member of the board of directors of the Roosmoor Community Services District. Dr. Barke is a co-founder and current board chair of Orange County Classical Academy, a free public charter school in Orange, California. Dr. Barke also serves as a reserve deputy and tactical physician for an Orange County law enforcement agency and as a commissioned officer in the US Army Reserve Medical Corps.

Dr. Barke co-authored *The Essential Diet Planning Kit* (with Godfrey Harris) in 2005. He has appeared multiple times on the *Dennis Prager Show*, the *Larry Elder Show* and on *Fox News Special Report* with Bret Baier.

He is married to his high school sweetheart, Mari Barke, currently the president of the Orange County (California) Board of Education. They have two adult children, both of whom live out of state.

ACKNOWLEDGMENTS

I WISH TO THANK MY WIFE of more than 30 years, **Mari Barke,** for her support, ideas and tolerance of my new social media activities. She is also the photographer who made the head shot of me that appears with my biography.

My daughter, **Allie Barke**, played an important role in helping with social media and creative content. She also designed our logo (shown at the end of this section) and maintains my website:

www.rxforliberty.com

Thank you also to my son, **Sam Barke,** for pushing me to stay in the fight with courage. Special appreciation goes to my business associaate, Dr. **Kenneth Cheng,** for allowing my passion to flourish and to the staff of **Personal Concierge Physicians** of Corona del Mar, California, for fielding calls to our offices — not all of which have been supportive!

I am grateful to **Dennis Prager** and **Larry Elder** for inviting me to share my views on their radio programs and for writing the book's Foreword (Prager) and saying those nice words on the book's back cover (Elder).

After the first edition was published, **Kirk Cameron** said this: "In a news-cycle when propaganda is more common than truth, Dr. Barke gives us researched medical facts and common sense about COVID-19 to help us thrive in the midst of an unprecedented national panic-demic." **Pastor Rob McCoy** of Godspeak Calvary Chapel, wrote: "The facts and truth contained in [Dr. Barke's book] have served to empower people to overcome fear and stand against tyranny of those who would use lies to take our freedom. His courage and wisdom have been a great blessing to me personally as well as those I serve." I am moved by these words and appreciate them deeply.

I was taught that there is no such thing as good writing, only good rewriting, so thank you to **Godfrey Harris** and The Americas Group's stable of specialists, especially editor **Art Detman,** without whose help this book would not be possible.

Thank you also to the *American Thinker* (www.americanthinker.com) for publishing most of these essays. Special appreciation to **Earick Ward** for co-authoring some of the pieces in the book and supporting my efforts to give our ideas a wider audience.

I must also acknowledge the work and contribution of **Simone Gold**, M.D., J.D., founder of America's Frontline Doctors. I am proud to be a member of this courageous organization empowering patients and physicians with independent, evidence-based medicine.

Finally, it has been quite a bumpy ride for the United States over the past few years. I have tried to put forward a common sense approach to dealing with Covid-19 issues and the societal upheaval that was in part spawned by the pandemic.

I have been called a hero by some and a quack by others — my home county is fractured by vehemently held opposing views like never before. I will continue to write and speak as time permits and express my thoughts on social media. I have been humbled to have been given an opportunity to have my voice heard across America and carried into foreign lands.

I am hopeful God will bless me with continued energy, strength and wisdom to lead with facts and science and help America.

I am looking forward to feedback from readers in the hopes of improving my message to better communicate common sense to close the widening gap between healthcare and politics.

Dr. Barke's blog can be accessed at
RxForLiberty.com

INDEX

A

Abbott Laboratory — 80
Abuse — 50
Accidents — 15
Addicts — 9
Adenovirus — 91, 92
Africans — 29
Age — 55, 97
Air — 40
Airplane — 63
Air Travel — 102
Alcohol — 40
Allergies — 22, 64
Alone — 47, 51
America — 111, 112
American Thinker, The — 110
America's founding principles — 65
America's Frontline Doctors — 111
Americas Group, The — 110
Americans
  History — 15
  Outpatients — 16
  Society — 86
  Way of Life — 50
  Youth — 14
Anarchy — 64
Animal studies — 96
Antibodies — 90, 100
Antibody Dependent Enhancement (ADE) — 93, 94
Anti-inflammatory — 44
Anti-bacterial — 22
Anxiety — 73
Assays, Imunofluorescence-based — 77
Automobile accidents — 83
Assisted Living Facilities — 14
Asthma — 22, 36
AstraZeneca — 91, 92
Asymptomatic — 31, 92
Ativan — 17
Author — 10, 107-108

Authorization — 106
Automobile — 15
Accidents — 30, 83
Azithromycin — 36

B

Bacteria — 20, 22
Bad apples — 64
Baire, Brett — 108
Bandanas (see Masks)
Barke, Allie — 109
Barke, Dr. Jeffrey Barke — 8, 10, 11, 12, 73, 107
Barke, Mari — 108, 109
Barke, Sam — 109
Barrington — 72-73
Bars — 44
Bartlett, Dr. Richard — 35
Beaches — 30
Beer — 49
Beder, A. — 29
Behavior — 24
Benzodiazepines — 17
Better Safe than Sorry — 87
Bhattacharya, Dr. Jay — 72
Bible — 7
Biography — 109
Birth — 21
Blood clots — 92, 96
Blood pressure — 87
Board of Education — 106
Bodies — 22, 23
Boosters — 53-60, 68
Brandon Dunes, Oregon — 77
Breakthroughs — 57-58, 68
Bubble Boy, The — 21, 23
Budesonide — 36
Bus seats — 23
Business shutdown — 48

C

California — 65
Cancel culture — 105
Cancer — 17, 31

113

Cardiologist — 17
Cardiovascular
    Disease — 23
CDC — 8, 20, 30, 43, 70,
    74, 99
Cells — 20, 21
Center for Disease
    Control and Preven-
    tion (See CDC)
Chainlink fence — 26
Change — 12
Charter schools — 108
Chemotherapy — 31
Cheng, Dr. Kenneth
    — 110
Chief Medical Advisor
    — 106
Children — 20, 23, 30,
    31, 48, 49, 73, 79, 84,
    101
Churchill, Winston
    — 18
Cities — 51
Civil disobedience — 74
Cleaning products
    — 22
Clinical trials — 55
Clothes — 24
C19/O — 67-74 (See also
    Omicron variant)
Coaches — 49
Coetzee, Angelique — 67
Coincidence — 42
Colonoscopies — 17
Common cold — 27
Common sense — 25,
    29, 30, 31, 32, 111, 112
Comorbidities — 73
Concierge — 105
Consensus — 74
Control — 67
Conventional Wisdom
    — 11, 12
Coronavirus — 9, 25-26
    (See also Covid)
Coronavirus Task Force
    — 16
Costs — 36
Cough — 27, 28, 29
County — 111
Courage — 9, 10, 18, 105
Courtesy — 62
Covid — 9, 11, 23, 105-106

Covid (Continued)
    Asymptomatic — 18
    Blacks — 99
    Cases — 33, 34, 42
    Deaths — 34, 36, 42,
        43, 44, 68, 104
    Decisions — 14
    Delta variant — 53
    Dollars and sense —
        54
    Dosage — 55
    Early treatment —
        35-36, 60, 71
    Edict — 25
    Cough — 27
    Fact — 12
    False information
        — 18
    Fatalities — 14-15
    Fatality rate — 45
    Fear — 17, 18, 43
    Fetuses — 101
    Focused Protection —
        72
    Food and Drug
        Administration —
        54-55
    Harm — 105
    Healthy — 43
    Hospitalizations — 34,
        35, 39, 45, 58
    Hospitals — 104
    ICU — 35
    Immunity — 106
    Infections — 35
    Laboratories — 15
    Masks — 25, 27
    Mild Symptoms
        — 43, 45, 58
    Minority groups — 99
    Mistakes — 15
    Myths — 12
    New Cases — 45
    C-19/O — 73
    Omicron variant
        — 67-74
    Particles — 28
    Positive — 37
    Post-vax infections —
        58
    Reactions — 14, 15
    Recovered — 54
    Risks — 30, 31, 32, 35

Covid (Continued)
  Size of virus — 25
  Size comparison — 26
  Spike — 40
  Spread — 17, 25
  Studies — 27
  Survival — 35, 38, 45, 57, 97
  Symptoms — 45, 105
  Testing — 18, 110
  Therapeutics — 34
  Threat — 12
  Transmission — 31
  Tools — 104
  Transmission — 92
  Treatments — 34, 104
  Vaccinated — 68
  Vaccination — 105
  Vaccine — 105, 106
  Young — 41, 105
Cuomo, Governor Andrew — 65
Curiosity — 106

D
Danish — 25
Data bank — 75
Data doubt — 79
Daughter — 107
Deaths — 30, 33, 42, 44, 45 (See also Covid-Fatalities)
  Vigils —48
Debate — 12
Decency — 63
Decision making — 87
Declaration — 72
Declaration of Independence — 11
Decorum — 62
Delta variant — 67
Dennis Prager Show — 108
Depression — 9, 73
Deputy — 108
Detman, Art — 110
Detroit — 36
Dexamethasone — 44
Diabetes — 23, 83
Dirt — 23
Disagreements — 65
Discounts— 49

Dissent — 11, 69, 106
Distance — 69
DNA — 96
Doctors — 18, 34 (See also Physicians, Surgeons)
  Politics — 38
  Treatments — 39, 44
Dogs — 64
Door handles — 23
Drinking — 87
Droplets — 27
Driving — 86
Drowning — 30, 43
Drugs — 9

E
E. Pluribus Unum — 65
Eating — 38
Echocardiograms — 17
Economics — 16
Economy — 33, 42, 85
Education — 9
Elder, Larry — 110
Elderly — 14
Elections — 108, 111
Emergencies — 67, 68, 70
Emergency Use Authorization — 71, 89
Emotional support animal — 64
Enemies — 69
England — 57
Essays — 12
Essential Diet Planning Kit, The — 108
Establishment — 105, 106
Ethiopia — 56
European — 29
Exercise — 22, 29 38
Experts — 54, 67, 72, 73, 74
F
Face covering (See Masks)
Face shield — 62
Fake news — 79, 81
Family — 9, 34
Fast food — 83
Fatality rate — 42, 43

(See also Deaths)
Fauci, Dr. Anthony
— 16, 25, 68, 70, 74,
106
FDA — 8, 70, 89
Fear — 24, 31, 67, 68, 70,
72, 74, 87, 88, 89, 106
Feedback — 112
Fence — 28
Fights — 50, 61
Filters — 26, 27
Financial incentives
— 54
Fingers — 28
Fitness — 22
Flattening the curve
— 42
Flights — 48
Flu (See Influenza)
Focused Protection — 72
Food Supply — 83
Foods — 86
*Fox News Special
Report* — 108
Frail — 14
Freedom of speech
— 11
Freedom of Choice
— 103
Freedoms — 61, 69
Friendship — 108
Fungus — 20
Futility — 65

G
Gain of Function
— 105, 106
Gas pumps — 23
Gastroenterologist
— 17
Genetic code — 90
Germs — 20, 21, 22, 23,
64
God — 8, 65, 111
Ghebreyesus, Director-
General Tedors
Adhanom — 56
Goebbels, Joseph — 69
Gold, Simone, M.D.,J.D.
— 16, 111
Government — 30, 62,
69, 70, 102

Grants — 71
Great Barrington
Declaration — 72
Great Depression — 48
Greek — 13
Grievances — 61
Gupta, Dr. Sunetra — 72
Gynecologist — 17

H
Hair — 26
Hand sanitizer — 20, 22
Hands — 28
Happiness — 61
Harm — 13, 14
Harris, Godfrey — 12,
108, 110
Harvard University — 72
Hayek, F. A. — 68
Hazmat — 62,63
Guy — 65, 66
Headaches— 28
Headlines — 32, 41, 42
Health — 19, 28, 32, 45
Politics — 31
Healthcare — 64, 105, 112
Heart
Disease — 83
Failure — 46
Heaven — 10
Henry Ford Health
Systems — 36-37
Hepatitis-C — 89
Herd immunity — 74
Hero — 111
Hiding — 62
Hippocrates — 13, 18
Hippocratic oath — 56
HIV — 89
Hoag Memorial
Hospital — 107
Holy — 8
Homes — 23
H1N1 virus — 85
Hong Kong Flu — 84
Hope — 89
Hose — 23
Hospice care — 46
Hospitals — 34, 45, 71
Houses of Worship
— 48
Hugs — 46
Hydroxychloroquine

— 36, 38, 44
Hyper-evaluation
— 64, 65

I

ICU — 23, 35
Ideas — 87
Immune compromised
— 55
Immune system — 19,
20, 21, 23, 31, 39, 90
Immunity — 54, 100
Impoverishment — 9
Inactivity — 9
Inaugural Address — 86
Income — 9
India — 95
Infection — 28, 29, 31,
38, 42, 43, 44
Infectious diseases — 22
Infectious Diseases Soci-
ety of America — 73
Inflammation — 45
Influenza — 14, 27, 45
Choice — 98-99
Efficacy — 97
History — 97
Risk — 98
Injury — 22
International Border
Crossing — 102
Internet — 37
Intestinal obstruction
— 18
Intestines — 22
Israel — 53, 58
Isolation — 31, 45

J

JAMA Pediatrics —
79-80
Jefferson, President
Thomas — 106
Jobs — 16
Johns Hopkins — 75
Johnson & Johnson —
71, 92,
Judaism — 7
Justice — 62, 64, 66

K

Kaiser-Permanente — 55

Kennedy, Robert F., Jr.
— 74
King George III — 11
Kisses — 46
Kulldorff, Dr. Martin
— 72

L

Lancet, The — 8, 106
Larry Elder Show — 63
Latin — 13
Law enforcement — 63,
107
Lawlessness — 62, 63
Learning disorders — 73
Lecturer — 10
Leftovers — 56
Letters — 16, 51
Liability —98
Liberty — 64, 69, 74, 88,
106, 112
Lies — 63, 69, 73
Life — 61, 64
Liquor — 49
"Live Free or Die" — 87
Living Life — 88, 89
Lockdowns — 8, 9
Loneliness — 48
Long Covid — 97
Looting — 61, 64, 65
Los Alamitos Unified
School District
— 108
Lotions — 22
Luck — 50
Lung Disease — 46

M

Magician — 67
Mainstream Media
— 88
Malaria — 90
Malice — 16
Mammograms — 17
Mandates — 74
Marines — 94
Market — 23
Masks — 18, 20, 25, 26,
27, 28, 29, 30, 32,
44, 61, 62, 69, 73, 74,
84, 85, 106
Airplanes — 48, 61-
66

117

Masks (Continued)
    Cloth — 26
    Efficacy — 63
    N110 — 26
    Police — 62
Massachusetts — 57
Math — 59
McDonald, M.D., Mark
    — 64
McCoy, Pastor Rob
    — 110
Meals — 46
Media — 12, 51, 54, 67,
    69, 75, 81, 88
Medic — 63
Medications — 72
Medical
    Establishment — 9
    Expertise — 10
    Father — 13
    Observations — 13
    School — 88
    Science — 13
    Treatments — 9
    Wisdom — 8
Medical Injury Com-
    pensation Reform
    Act — (see MICRA)
Medications
    Recommended — 40
Messenger RNA
    (mRNA) — 90
Meter — 29
Michigan — 65
MICRA — 99
Microbes — 22
Microbiome — 22
Microns — 25-26
Minds — 12
Minneapolis — 62
Mistakes — 9
Mob — 62
Moderna — 71, 90, 92
Moisture — 28
Mold — 20
Monoclonal antibodies
    — 60
Mosque — 48
Mosquitoes — 26
Motorycles — 86
Mountain climbing
    — 86

Muller, M.D., Ph.D.
    William — 80
Mouths — 28, 29
Mucus — 28
Multi-ethnic — 63
Murder — 62
Mutations — 104
Myocarditis — 101

N
Nasal rinse — 71
Nasopharynx — 79
Nation — 66
National Anthem — 87
National Childhood
    Vaccine Injury Act
    (See NCVIA)
National Guard —65
National Institutes of
    Health — 70
Natural immunity
    — 68
Nazis — 63, 69
NCVIA — 98
*New England Journal
    of Medicine* — 8, 74,
    79
New Hampshire — 87
New York — 85
New York City 63
New York Times 1619
    Project, The — 63
Newport Beach — 11,
    107, 110
News — 34, 44
Newsom, Governor
    Gavin — 25, 65
Newspapers — 41
NIH (See National
    Institutes of Health)
N95 — 62
New York — 65
Nobel Prize — 68
Non-essentials — 35
Nose — 28
Notes — 52
Novavax — 93
Nurses — 28, 29
Nursing homes — 14,
    46, 47, 48 (See also
    Assisted Living
    Facilities)
Nutraceuticals — 70

O
Oath — 13
Offices — 23
Officials — 18
Omicron variant — 67, 68
One-size-fits-all — 69
Operation — 28
Opinions — 11
Optimism — 89
Orange County — 30
Orange County Board
    of Education — 108
Orange County
    Classical
    Academy — 108
Orange County
    Medical
    Associa tion — 107
Orange County
    Sheriff's
    Department — 76
Orange, CA — 108
Overweight — 23
Oxford University — 72
Oxygen — 28, 29

P
Pandemic — 15, 31,
    32, 47, 105
Pap smears — 17
Parents — 59
Parks — 30
Particles — 29 (see
    also Coronavirus)
Passenger — 61, 63
Pathogens — 20
Pathways to
    Independence
    — 108
Patient — 16, 28, 75,
    78, 79, 88
    Doctor Visits — 17
    Impact of virus — 16
Peace —62
Personal Care
    Physicians — 110
Petitions — 61
Pfizer — 53, 71, 90, 100
    Efficacy – 55
Pharma — 53, 72
Photographer — 107
Physicians — 7, 8, 10,
    11, 13, 34, 41

(See also Doctors,
    Surgeons)
Police — 64, 65
Politics — 10, 27, 105,
    112
    Leaders — 11
    Left — 63
    Pressure — 32
Politicians — 13, 30, 64
Poor — 56
Pores — 25
Post Traumatic
    Power — 89
Povidone iodine — 71
Prager, Dennis — 10,
    34, 107
Prayer — 49
Precautions — 88
Preparations — 88
Prescriptions — 17
President — 64
Price — 9
Primary care — 88
Procedures — 35
Profits — 72
Property — 64
Propaganda — 74
Protections — 31, 88
Proteins — 91
Protests — 33, 42, 43,
    44, 62, 63, 64
Protocols — 36
Psychiatrist — 17
Psychological
    pathology — 64
PTSD — 30
Public health — 11
Public opinion — 30
Public Policy Decisions
    — 82
Pulmonary
    inflammation — 36
Pulse rate — 29
Purchased products
    — 23

Q
Quack — 111
Questions — 52, 65
Queues — 49

R

Rabbis — 8, 10
Racism — 63, 64
Radicals — 64
Radio — 10
Rally — 49
Real world — 55
Re-infection — 28
Re-opening — 42, 44
Reactionary — 65
Readers — 12, 112
Redfield, Dr. Robert — 25
Regeneron — 58, 60
Reimbursement — 15
Remdesivir — 36
Repurposed medications — 60
Research — 25
Restaurants — 44
Retail stores — 44
Retests (See Tests)
Revolutionary — 65
Ribonucleic acid (See RNA)
Riots — 33, 42, 44, 63, 65
Risks — 15
Risky — 73
RNA — 79, 90-91
Road to Serfdom, The — 68
Rossmoor— 106
Rude — 62
RxForLiberty.com — 109, 110, 111

S

Safety — 29
Sanitization — 23
SARS-CoV-2 — 79, 81, 95, 102
School Board — 50
Schools — 15, 19, 30,
  Classrooms — 20
  Closing — 42
  Hand sanitizers — 20
  Open — 15
  Plans — 19
  Re-openings — 19
  Sanitizing spray — 19

Schools (Continued)
  Staffs — 20
  Shutdowns — 43, 48
Science — 10, 27, 29, 32, 70, 74, 105
Scientists — 67, 73, 89
Seinfeld, Jerry — 21
Self-esteem — 64
Seniors — 99
Sensation-mongering — 32
Separation (See Social Distancing)
Sepsis — 18
Shaming — 65
Shaw, George Bernard — 12
Shopping — 48
Shortness of breath — 59
Shower — 24
Shutdown — 45
Sick — 32, 45
Single path — 69
Six feet — 29
65-years — 32, 44
Size — 25-32
Skepticism — 106
Skydive — 86
Slavery — 86
Sleep — 40
Sleight-of-hand — 67
Slogans — 53
Smell — 67
Sneeze — 27, 29
Soaps — 22
Social distancing — 20, 29, 32, 44
Social media —62, 109, 111
Society — 32, 45
Socio-political agenda — 64
Sore muscles — 67
Soul — 50
South Africa — 67
Speaking — 51
Special ops — 21
Specialists — 20
Spikes — 43, 44
Spittle — 27, 28
Stanford University — 72

Statistic — 33
Stay-at-home — 17
Sterilization — 23
Steroids — 36, 44
Store-owners — 64
    (See also
    Retail Stores)
Studies — 31, 37
Sugar — 40, 63
Suicide — 9, 48, 73, 83-84
Sundays — 46
Sunshine — 40
Supplements — 38
Surge — 35
Surgeons — 28, 29,
    30 (See also Doctors,
    Physicians)
Surgeries — 28, 35
Surgical gloves — 62
Survival — 21
SWAT — 63
Symptoms — 45, 67
Synagogue — 48
Syphilis — 99
Systematic — 64
Systemic — 64

T
Talk shows — 10
Talking — 28, 29
Talisman — 27
Talmud — 7, 8
Taste — 67
T-cells — 100
Teachers — 31
Teasers — 42
Teens — 57
Television — 41
Temperatures — 92
Testing — 33, 45,
Tests — 55, 75, 78
    Abbot Laboratory —
        80
    Accuracy — 76
    Bogus results — 75
    Ghost — 76
    Positive — 75
    Results — 76
    Trump, Donald — 77
    Negative results — 78
    Positive Results
        — 78, 79
    Sofia rapid nasal — 77

Tests (Continued)
    Validity — 76
    Viral Load — 79
    Television — 21
Texas — 36
Theme Parks — 102
Third shot — 55
Third-world countries
    — 92
Tiredness — 67
Tissue — 28
Totalitarianism — 69
Touch — 28, 47
Travel — 68
Treatment — 34
Trick — 67
Trump, Melania — 77
Trump, President
    Donald J. — 16, 36,
    77
Trust — 74
Truth — 69, 71, 74
Truth — 63
Tutoring — 57, 59, 51
Tyranny — 74

U
Ugly — 32
Unalienable rights — 61
Unemployment — 16
Unions — 30
United States — 8
    (See also America)
University of
    California Irvine
    — 107
University of
    Southern
    California — 107
Unvaccinated — 54, 58
U.S. Public Health
    Service — 99
Utopia — 65

V
Vaccinations — 18, 67,
    71, 89
Vaccine Adverse
    Event Reporting
    System (VAERS) — 59
Vaccines — 69, 74, 89, 91,
    92

Vaccines (Continued)
  Author's Perspective
    — 59, 60
  Booster—53-60
  Choice — 98
  Claims — 53
  Consent — 103
  Controlled Study
    — 95
  Dosage — 53
  Effectiveness — 53, 55
  Efficacy — 92, 94
  Failure — 59, 102
  Government — 89
  Guillain-Barré Syn-
    drome — 100
  Healthcare workers
    — 94
  Impact of Age — 96
  Ineffective — 74
  Injuries — 68
  Liability — 97.
  Mandates — 101, 103
  Manufacturing — 98
  Opposition — 102
  Paradoxical Reaction
    — 93
  Passport — 103
  Protection — 55
  Questions — 103
  Safety — 101
  Savior — 104
  Shortages — 56
  Storage — 92
  Side effects — 56
  Survival Rate — 96
  Technology — 57
  Temperature — 92
  Testing — 94
  Variant — 67
Vaccinated — 58
VAERS — 59, 102
Valium — 17
Value for money — 46
Van Kerkhove, M.D.,
  Maria — 31
Videos — 44
Vigilance — 106
Violence — 65
Viral load — 79
Viruses — 20, 23, 69 (See
  also Covid, Influenza)
Visitations — 47

Visits — 31, 46
Vitamin C — 40
Vitamin D3 — 40, 71
Votes — 64
Vulnerable — 32, 60,
  72, 88,

W
Walk — 28
War —50
Ward, Earick — 110
Warp Speed — 89
Washington, D.C.
  — 61, 89
Weight — 37, 83
White — 63
White House, The
  — 16
Whitmer, Governor
  Gretchen — 65
WHO — 14, 26, 31,
  56, 74
Wife — 108, 109
Winning — 65
Wisdom — 10
Woodstock — 85
Work — 84
World Health Organ-
  ization (See WHO)
World Wars — 86
Worship — 48
Wounds — 28
Wuhan Lab —105, 106

X
Xanax — 17

Y
Yes — 64
Young — 32
Youth — 44

Z
Z-pack — 44
Zelenko, Dr. Vladimir
  — 34, 36, 38
Zinc — 36, 39, 40, 44